HYPER-PERSONALIZED
VENUES

HYPER-PERSONALIZED VENUES

A CEO's Guide to AI, Privacy, and World Models

Companion to the reference standard

The World Model
Governed AI and Hyper-Personalization in Physical Spaces

MARIS J. ENSING

Copyright © 2026 by Maris J. Ensing

No vendor, product, or implementation is endorsed by this publication.

This publication provides information about the subject matter covered. It is distributed with the understanding that the publisher is not engaged in rendering legal, engineering, or other professional services. If expert assistance is required, the services of a competent professional should be sought.

WorldModel™ and related marks are trademarks of Creative Mad Systems, Inc. All other trademarks are the property of their respective owners.

No part of this publication may be reproduced, stored in a retrieval system, or transmitted in any form without prior written permission.

Companion resources and framework updates: worldmodel.global.

Hyper-Personalized Venues, Maris J. Ensing —1st ed.

The first paragraph must earn the next

If you lead a venue, your most dangerous competitor is rarely the one with the newest building or the loudest marketing budget. It is the operator who makes the day feel easy—not by dumbing it down but by arranging hundreds of small moments to align and then letting that alignment compound.

Ease is not a feature you install. It is the outward sign of an inward architecture coherent enough to hold up under load, on a bad day, with tired families, late groups, crowded corridors, and staff already stretched. Guests experience that coherence as modern. Staff experience it as calm. Finance sees it as higher performance per headcount.

◆ *Coherence is what happens when a place behaves like one place.*

Ticketing, wayfinding, content, queues, accessibility posture, and support moments align around the same question: what is happening now and what should happen next?

Most guests will never analyze why a day feels different. They will simply prefer it, and that preference hardens into habit, which becomes market share.

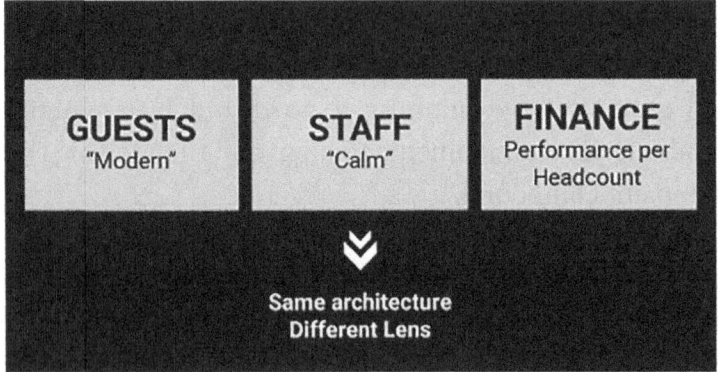

SCENE: SATURDAY, 2:17 P.M.

A family pauses under the main sign because the arrows contradict the map, the map contradicts the app, and the app asks them to choose a language again. They are not angry. They are simply doing math, deciding which parts of the day to skip. That is what fragmentation costs. It quietly steals attention from the experience, turning an afternoon into a sequence of negotiations.

Phones trained people to expect continuity and immediate usefulness. The next category of physical experience is a place that feels as helpful as the device in your pocket while behaving more respectfully than the platforms that trained you.

Coherence compounds, and friction compounds, too: small contradictions multiply into questions, questions become queues, queues become stress, and stress suppresses dwell, spend, learning, and return intent.

What this brief is meant to do

This brief is a companion to the reference standard, **The World Model: Governed AI and Hyper-Personalization in Physical Spaces**, which outlines the full architecture, governance patterns, procurement language, and acceptance criteria.

This brief is intended for CEOs, boards, owners, and senior designers. It is intentionally non-technical and aims to help leadership teams identify what has changed, understand what becomes possible, and demand the right outcomes in procurement and operations.

◆ | *Modern is behavior, not hardware.*

CONTENTS

The first paragraph must earn the next 5
Executive Summary..11

PART I: WHAT CHANGED

1. Phones Changed Expectations 17
2. Experience Debt and Fragmented Truth 19
3. Compute Nodes vs. Black Boxes 21

PART II: THE TECHNOLOGY THAT EXISTS TODAY

4. The Personalization Stack: A Complete System..............................27
5. The Technology Stack Diagram 33
6. The Three Modes of AI Delivery 37
7. Personal Delivery and Accessibility..43
8. Anonymous Operation and Privacy...49
9. Virtual Docents and What They Must Never Do...........................55
10. Lifecycle Automation and Verification..57

PART III: THE LADDER

11. Stage 0, Stage 1, Stage 2 61
12. Decision Guide: Which Stage is Right? 67

PART IV: WORLDMODEL AND GOVERNANCE

13. The Discovery: When Personalization Meets Scale.........................71
14. WorldModel In Plain Language 75
15. Governance, Constitution, and Decision Gates 81

PART V: WHY NOT
16. When This Approach Is Wrong 87

PART VI: DEPLOYMENT BY VERTICAL
17. Patterns Across Verticals 93

PART VII: BOARD-LEVEL DILIGENCE
18. Why This Matters At Board Level 105
19. What To Demand In Procurement 109
20. What To Demand In Operations 113
21. Questions For Partnership Conversations 115

PART VIII: THE CATCH-UP PREMIUM
22. What Happens If You Do Nothing 119

APPENDIX A
Executive Glossary 125
Sources 129
Companion Resources 131
A Note from The Author 133
About the Author 137
Other References 139

Executive Summary

What you are deciding: whether your venue's technology investment compounds into coherence or fragments into operating drag.

What changed: phone-era expectations became the benchmark, including continuity, translation, accessibility, and immediate usefulness.

What this brief gives you: the Stage Ladder (0, 1, 2), the non-negotiables (privacy posture, governance, verification), and the questions that force evidence, not claims.

Most venues cannot deliver this.

Over years of sensible additions—ticketing systems, apps, queue tools, interactives, and dashboards—they have accumulated what this book calls experience debt. Each system solved a local problem but created another model of reality. Now, signage says one thing, the app says another, and staff say a third. Visitors feel these contradictions immediately. They cannot afford to be wrong when they are tired, rushed, or navigating in an unfamiliar language.

Smartphones changed what people expect. Through daily use, visitors learned that systems should remember what they already said, that assistance should arrive when needed, and that translation is a default condition of modern life. Physical venues are now measured against that standard. The benchmark moved, and it is not moving back.

◆ *Coherence is what happens when a place behaves like one place.*

This book argues that coherence is not a feature you install. Instead, it is an architecture you build. The technology exists today: language selection and accessibility at every touchpoint, depth adaptation to match visitor intent, privacy-forward recognition that maintains continuity without surveillance, and AI-powered guidance bounded by institutional knowledge. These capabilities are deployed in museums, brand centers, and attractions now.

They work.

In this book, "proven" means production-deployed and repeatable under peak-day conditions. If a vendor cannot demonstrate the capability under load, with measurable tolerances, documented failure behavior, and auditable logs, the capability is not proven.

That is why this brief includes specific acceptance questions you can require in procurement.

The question for leadership is not whether to invest in technology. Every venue does. The question is whether that investment compounds or fragments. Fragmented investment buys screens and apps that solve immediate problems while making the next upgrade harder. Compounding investment builds on a foundation that enables growth without reconstruction.

Not every venue needs the same capability. A single gallery may excel with a carefully designed static experience. A museum with multiple exhibits benefits from journey continuity: set your language once, and the venue honors it throughout. Only theme parks, cruise ships, airports, and districts require full venue-scale coordination. The book provides a staged framework to help the board determine what is right for their situation and avoid buying complexity they do not need.

The next decade will see entertainment destinations, cultural districts, and smart cities built at unprecedented scale—many in regions creating entirely new visitor economies. These developments cannot retrofit coherence later.

They require it from the ground up. Operators who define what visitors expect will set the standard. Operators who follow will pay a catch-up premium that compounds each year.

◆ | *Visitors remember how the day felt.*

For boards and leadership teams, this book provides a shared vocabulary to evaluate investments, specific questions to ask in procurement, and operational metrics that reveal whether coherence is improving or degrading. The companion reference standard offers the technical depth for those who need it. But the strategic decision starts here: determine which stage is right for your venue, demand an architecture that can evolve, and build the foundation before your competitors do.

PART I

WHAT CHANGED

1

Phones Changed Expectations

Phones taught people to expect continuity, instant translation, and guidance that adapts. Through repetition, people learned that systems should remember what they already told them, that assistance should arrive when needed, and that translation is a default condition of modern life.

Physical venues are now measured against that experience. Guests describe the gap as wasted time, repetition, and confusion—a day that feels harder than it should because the venue acts as if context must be rebuilt from scratch at every touchpoint.

The benchmark moved, and it is not moving back.

When expectations shift, point upgrades stop working because the failure is not confined to a single touchpoint. The failure is that the venue keeps rebuilding context from scratch while visitors expect continuity. Incremental upgrades often add another interface on top of inconsistent underlying truth.

This is why incremental upgrades disappoint. A new screen does not fix a hard day. A new app does not fix a fragmented reality. Behavior changes when the operating model changes.

◆ | *The mature posture: design for phone-era expectations without requiring a phone*

This matters because phones are optional in real life: batteries die, signal fails, accessibility tools vary, and not every guest should be forced into BYOD. Phone-era expectations are about continuity and usefulness, not about forcing a device.

A venue that holds this stance can naturally incorporate future upgrades—on-device agents, negotiated preferences, minimal proofs—because the architecture treats the phone as a privacy boundary rather than a data funnel.

A privacy-boundary posture keeps preferences with the visitor and limits what the venue must know, for the shortest time needed. That reduces exposure, avoids identity hoarding, and keeps future upgrades compatible by design.

2

Experience Debt and Fragmented Truth

Experience debt accumulates through sensible additions: ticketing, apps, queue tools, interactives, show control, dashboards, and one-off integrations built to meet deadlines. Each solves a local problem but creates another model of reality in which "truth" can diverge.

Over time, the venue becomes a federation of subsystems that disagree about what is happening now. That disagreement slows change, forces staff to act as translators, and turns upgrades into one-off events rather than improvements that compound.

Experience debt shows up when operational promises disagree: timed entry says one thing, the live queue condition says another, the map routes people through a closure, and staff improvise exceptions one family at a time. Guests experience it as friction. Operations experiences it as manual intervention.

In a busy venue, contradictions do not stay local. A contradiction creates a question. A question creates a staff interruption or a guest stop. Those micro-stops accumulate into a slow-moving front, which becomes a queue. Queues raise stress, and stress reduces dwell time, spend, learning retention, and return intent.

The cure is not another interface. Another interface only rearranges the disagreement.

The cure is shared state: one definition of "what is happening now," exposed to every touchpoint. It removes contradictions at the source, reduces staff translation load, and makes upgrades safer because changes are made once, verified once, and reflected everywhere, without re-integration across the venue.

SCENE: A CORRIDOR CLOSES

A corridor is blocked. The map, signage, staff tool, and personal delivery update to the same new route. Guests see one answer, not four. Staff stop translating contradictions and start helping with exceptions.

3

Compute Nodes vs. Black Boxes

A modern venue is a decade-scale asset that evolves over time with seasons, expansions, new exhibits, new compliance pressures, and new audience expectations. It cannot be managed on a two-year product cycle without paying a catch-up premium that compounds with each replacement.

The trap is familiar. A box reaches end-of-life. Spares run dry. Firmware maintenance stops. The replacement triggers reintegration across the venue. A "small upgrade" becomes a capital event.

Commodity Compute changed that dynamic permanently.

Solid-state computers are built by the millions and supported by deep supply chains. That ecosystem provides predictable replacement paths and well-understood failure modes. You can swap a node instead of replacing a subsystem.

A Compute-node architecture uses that ecosystem intentionally.

Instead of proprietary boxes, the venue runs on small nodes sized to the job. A light-duty node handles simple playback. A heavier node handles richer media and local inference. You provision the right Compute where it matters and retain the option to change that decision later. Upgrading the system to add additional capabilities later is simple, not a major decision.

What a Compute-node architecture consolidates

Traditional AV requires separate proprietary appliances for media playback, show control, audio processing, and display control. A Compute-node architecture consolidates these into general-purpose, non-proprietary hardware. This is a category change that determines whether the venue can evolve or must be rebuilt.

- ◇ **Lock-in failure modes to recognize**
- ◇ Vendor end-of-life forces rebuild cycles
- ◇ Closed toolchains limit auditability
- ◇ Supply chain shocks become redesign events
- ◇ Opaque licensing becomes an operational tax
- ◇ Tightly coupled integrations make changes risky
- ◇ Capability upgrades require infrastructure replacement

◆ *One question cuts through the fog: what happens when you want more capability next year?*

If the answer is "replace the box," you are buying a dead end. If the answer is "upgrade nodes where needed," you are buying evolution.

— PART II —

THE TECHNOLOGY THAT EXISTS TODAY

4

The Personalization Stack: A Complete System

Before discussing venue-scale coordination, note that the core personalization technologies—recognition, content delivery, virtual docents, and personal channel—are fully functional systems that deliver value independently.

They do not require WorldModel™.
They do not require venue-scale infrastructure.
They work.

The personalization stack is deployed in museums, brand centers, and visitor attractions today. Visitors receive content in their language, at their preferred depth, on their own devices. The technology is proven.

PROOF PACK: WHAT "PROVEN" MEANS

Acceptance rule: if any item below cannot be evidenced in your environment, the capability is not approved for autonomy or scaling.

A vendor is "proven" when they can provide evidence you can audit:

1) *Evidence model: what is logged, what is retained, what is deleted, and on what schedule.*

2) *Privacy posture: default anonymous mode works fully, and opt-in continuity is explicitly bounded.*

3) *Outcome verification: commands are not outcomes, and outcomes are measured, evaluated, and recorded with a pass/fail outcome.*

4) *Degraded-mode behavior: the system stays coherent under load, partial outages, and RF congestion under a defined peak-load test.*

5) *Operations discipline: rollback, change control, and drift detection exist before autonomy is granted.*

6) *No-identity acceptance: the core experience passes with no app, no login, and no identity, and the venue can show the acceptance test result.*

7) *Timing and synchronization: the system publishes sync tolerances and end-to-end timing budgets, and it passes them under peak load and degraded operation.*

Stage 2 is introduced by limiting scope, not by relaxing safeguards. If governance, consent boundaries, or verification are deferred, the capability is not ready.

What works at Stage 0 (exhibit-scale)

- ◇ Language selection and delivery at each touchpoint
- ◇ Depth adaptation (Streaker/Stroller/Student)
- ◇ Accessibility modes (captions, audio description, sign language)
- ◇ Privacy-forward recognition (session continuity)
- ◇ Virtual docent conversations bounded by Body of Knowledge
- ◇ Personal Channel delivery to visitor devices
- ◇ Offline-first operation for reliability

What works at Stage 1 (multi-exhibit)

- ◇ All Stage 0 capabilities, plus:
- ◇ Preference persistence across exhibits without re-asking
- ◇ Anonymous session continuity across a journey
- ◇ Coherent experience across multiple touchpoints
- ◇ No venue-scale infrastructure required

Stage 0 and Stage 1 are complete operating states, not steps toward anything else.

A museum that offers language selection, depth modes, and accessibility at each exhibit delivers genuine personalization value. Many successful venues operate at Stage 0 or Stage 1 and should feel no pressure to pursue higher stages.

Stage 1 is often the right long-term target. Journey continuity without identity hoarding, coherent language and accessibility across exhibits, and anonymous session recognition deliver most of the personalization value with manageable complexity. Many venues should operate at Stage 1 indefinitely.

What works at Stage 2 (venue-scale coordination)

At Stage 2, the venue behaves like one coordinated system, not a collection of well-designed exhibits. Personalization remains valuable, but Stage 2 adds real-time orchestration across touchpoints, operations, and constraints.

Stage 2 enables:

- Journey continuity across the whole site, not just across a cluster of exhibits.
- Real-time adaptation to operational state, including closures, delays, crowding, incidents, and capacity limits.
- Coordinated guidance and messaging across signage, kiosks, mobile, audio, staff tools, and show systems, so the venue does not contradict itself.
- Policy-aware personalization, meaning accessibility, safety, and brand constraints override preference when needed.
- Operational integrations that turn personalization into performance, including ticketing, scheduling, queue systems, membership or CRM, maintenance, and incident workflows.

Stage 2 requires:

- Venue-scale infrastructure, including reliable network coverage, venue telemetry, and sufficient on-site Compute to keep experiences responsive and resilient.
- A shared operational state model for the venue, so all touchpoints can coordinate.
- Formal governance, because the system can affect safety, access, routing, and service decisions, not just what content is shown.

Stage 2 Entry Gate (Go / No-Go)

Go only if all are true:

1) Shared operational state is authoritative.
   Touchpoints do not contradict each other because they reference the same operational truth.

2) Governance exists as a formal layer.
   Decision rights, constraints, and refusal behavior are defined because decisions affect routing, safety, access, and service.

3) Outcome verification is first-class.
   The system can prove outcomes under load. Missing verification forces conservative refusal, not silent failure.

4) Incremental means bounded scope.
   Start with fewer zones, fewer decision types, and fewer channels, but keep full governance, full outcome verification, and intact consent boundaries from day one.

If any item is missing, remain at Stage 0 or Stage 1 until it is satisfied.

Do not assume Stage 2 is the goal. Stage 2 adds capability but also increases complexity and governance requirements. Pursue it only when your venue's scale genuinely requires it.

5

The Technology Stack Diagram

The technology stack is layered. Each layer is complete and functional. Higher layers add capabilities for larger or more complex venues, but they are not required for the layers below to work.

◆ THE TECHNOLOGY STACK ◆
Executive View

Each layer is complete. Higher layers add capability, not requirement.

LAYER 3: GOVERNANCE (Stage 2 only)

Cognitive Governance Layer (CGL™)	**Constitution**	**Value System**
Decision gate that permits, modifies, or rejects actions	Explicit constraints: safety, accessibility, privacy, fairness	Safety over throughput, accessibility over convenience, trust over optimization

Added when: System makes consequential decisions affecting visitor flow, queue fairness, access, or safety at scale

↓

LAYER 2: VENUE ORCHESTRATION (Stage 2 only)

WorldModel™	**Venue Concierge Interface**	**Cross-Zone Coordination**
Shared operational truth: what is open, full, loud, safe	Translates visitor intent into outcomes	Flow balancing, queue shaping, scheduling

Added when: Multiple systems must agree on venue reality

↓

LAYER 1: BASE PERSONALIZATION STACK
★ Complete System - Works independently ★

Recognition Layer	**Content Delivery Engine**	**Virtual Docent Layer**
Privacy-forward session continuity	Language, depth, interest, accessibility	Conversational guidance from Body of Knowledge
Personal Channel	**Body of Knowledge**	**Lifecycle Automation**
Synchronized delivery: audio, captions, sign language	Curated, institution-controlled, source material	Occupancy-aware verification, drift detection

Deployed in museums, brand centers, and attractions today.
Does not require WorldModel™ or CGL™

↓

LAYER 0: COMPUTE SUBSTRATE (Foundation)
Non-proprietary nodes, offline-first, verifiable
Enables all layers. Prevents lock-in.

WHAT EACH STAGE REQUIRES

Stage	Layers Required	Governance
Stage 0	Layer 0 + Layer 1	Body of Knowledge only
Stage 1	Layer 0 + Layer 1	Body of Knowledge only
Stage 2	Layer 0 + Layer 1 + Layer 2 + Layer 3	CGL + Constitution + Value System

◆ **Layer 1 is a complete system.** *You can deploy, operate, and deliver genuine personalization value without Layers 2 or 3.*

Full technical diagram: worldmodel.global

6

The Three Modes of AI Delivery

When AI generates content for visitors, it must stay within institutional boundaries. Unbounded AI that invents facts or develops its own editorial voice is not permitted in public-facing systems.

Three delivery modes offer different balances of control and flexibility.

1. **Curated Library Delivery:**
   All outputs are pre-approved and scripted. AI may select, sequence, and route only from a fixed, approved content library. AI generates no new text, images, audio, or video.

   Use when:
   Regulatory or legal requirements require exact wording. Brand voice must be identical across visitors.

   What AI may do:
   Choose the correct approved variant by context, including language preference, accessibility mode, and recognition signals, then deliver it through the venue's media channels.

   What AI must not do:
   Create novel wording, invent facts, improvise answers, or synthesize new media.

2. **Institution-Governed BoK Delivery:**
   AI generates and adapts outputs only from the institution's approved Body of Knowledge and approved asset set. The BoK defines what the system can say, show, play, and recommend.

Use when:
Visitors need flexible depth, pacing, or modality (short, detailed, kid-friendly, expert-level, multilingual, accessible), and questions vary while topics remain bounded.

What AI may do (governed):

- Generate explanations, summaries, translations, and Q&A responses grounded only in approved sources.
- Adapt delivery to the visitor context, including language and accessibility requirements (for example, captions, audio description, simplified language), without disclosing identity.
- Select, assemble, and sequence approved media components into new combinations when permitted, to tailor the experience to context and intent.
- Produce multiple depth layers, and switch layers on demand, without changing the core facts.

Critical principle:
The BoK belongs to the institution, not the AI vendor. The institution controls sources, boundaries, voice, update cadence, and what topics are out of scope.

What AI must not do:
Use external sources by default, import its own "background knowledge" as authority, or generate content outside approved boundaries.

3. **Hybrid Delivery:**
   Curated headlines with BoK-generated depth and optional adaptive layers. The headline, key claims, safety language, and primary call-to-action are always scripted. Optional depth is generated from the institution's BoK and may be supplemented by approved media assembly.

   Use when:
   Consistency matters for the main message, but depth should adapt to visitors' interests, available time, and modality needs.

   Controls:

   - Scripted spine never changes.
   - Generative layers must remain inside BoK boundaries, and must be suppressible to revert to curated-only mode.
   - If the system can assemble new combinations, it must do so only from approved components and approved templates.

What all three modes prevent: Unbounded generation—when AI invents facts, improvises beyond approved sources, or develops independent opinions.

Why this matters for procurement: "We want AI personalization" is not a specification. "We want hybrid delivery with curated headlines and BoK-constrained depth" is a specification.

7

Personal Delivery and Accessibility

A personal delivery layer sends multilingual audio, captions, and accessibility formats to visitor phones and assistive devices, changing the economics of inclusion by scaling without turning every accommodation into a special request.

A browser-based guide reduces friction because visitors are tired of apps. Scan-and-go experiences often outperform app-first strategies.

Streaker / Stroller / Student: Three depth modes match visitor intent. Streaker wants the headline—fifteen seconds. Stroller wants context—30-45 seconds. Student wants full depth—unlimited, within reason. The visitor chooses; the venue adapts.

This matters operationally. Without depth modes, you either overload everyone with long explanations or you force staff to become the translator and the curator. Personal delivery lets the venue publish one canonical message, then adapt duration and format to the visitor's attention, language, and accessibility needs. The challenge is reliability under load, not content authoring.

Why synchronization matters

Synchronization is what makes personal delivery credible: captions must align with speech, audio description must align with visuals, and group moments must not drift.

Frame-accurate sync in dense RF environments is a genuine engineering challenge. Many vendors claim Personal Channel capability but cannot deliver reliable sync under load. Require measurable sync tolerances under load. Note that visitor BYOD selections will affect this—and that frame level accuracy is not always needed, and not always meaningful.

Why offline-first matters

Offline-first is how inclusion survives peak days: if the network degrades, the accessible path still works.

Venues have dead zones and interference from the crowd. Critical paths must run locally. A system that requires constant internet connectivity will fail during the busiest periods.

Accessibility is not negotiable. Multilingual delivery and accessibility are normal human conditions, not edge cases.

Inclusion Baseline (IDEA) and Operational Sustainability (ESG)

IDEA and ESG are non-negotiables you specify up front, commission, and keep true through upgrades.

A coherent venue treats IDEA and ESG as design constraints, not optional features.

IDEA by design (Inclusivity, Diversity, Equity, Accessibility):

The accessible path is the primary path. It remains usable in degraded modes, incident modes, and peak-load days.

Minimum baseline outcomes:

- Hearing support: captions for prerecorded content, assistive listening support, and clear signaling that these aids exist.

- Hearing aid compatibility: personal audio delivery that works cleanly with hearing aids, earbuds, and assistive receivers, without requiring special staff intervention.

- Vision support: audio description where relevant, screen-reader compatible digital delivery, tactile outputs where physical labeling is used, and Braille where appropriate.

- Non-visual navigation (blind and low-vision): a first-class guidance layer that enables independent movement without identity, designed to be dynamic, tied to operational truth, and safety constrained.

- Language dignity: multilingual delivery is not a translation afterthought. It includes captions, and it respects the correct sign language for the audience where sign-language interpretation is provided.

- Sensory safety: comfort modes, predictable pacing, and reduced cognitive load options, including neurodiversity-aware choices, without degrading the experience for everyone else.

- Acoustic intelligibility: sound is the most underestimated inclusion layer. A space can be accessible on paper and inaccessible in practice if speech is unintelligible due to reverberation, competing sources, or poor signal-to-noise. Visitors listening in a second language need better clarity, not louder volume. Specify intelligibility standards, not just equipment lists.

Personalization that stays respectful:

Personalization is the venue adapting to the visitor's chosen intent, not the venue guessing who they are.

- Depth modes: short headline, medium context, and full depth, chosen by the visitor.
- Interest support: preferences can persist across exhibits so people are not re-asked at every touchpoint.
- Age support: provide age-appropriate modes and entitlements through choice and verification, not inference.

SCENE: INCLUSION AT SPEED

A family uses captions. A blind visitor uses audio description. A school group shares a synchronized moment. If captions drift, or audio description lags the visuals, the experience fails.

Synchronization is not polish. It is the minimum for dignity at scale.

ESG by design (Environmental, Social, Governance):

Treat energy, thermal, network load, and Compute as measurable operating constraints.

- Compute efficiency: provision the right Compute, and prove it under peak attendance.
- Longevity: design for modular replacement and upgrades, with clear support life and end-of-life pathways.
- Responsible procurement: require vendor transparency on update policy, security posture, and serviceability.
- Governance in operation: assign accountable owners, maintain change control, and keep a small set of measurable metrics visible to leadership.

8

Anonymous Operation and Privacy

Anonymous operations improve flow, comfort, and guidance without creating dossiers. Identity is reserved for opt-in programs with a clear value exchange.

Three varieties of anonymous recognition

1. Session continuity: Preferences persist throughout a visit. A token maintains state; it is forgotten when the session ends.
2. Anonymous return recognition: A returning visitor is recognized without being named. "Welcome back" without surveillance.
3. Pattern recognition: Flow patterns and occupancy density measured at the place level, not the person level.

The drift risk

Anonymous systems can drift toward identification by accident:

- ◇ Tokens retained for too long could become shadow identities
- ◇ Multiple correlated data points become identification
- ◇ Biometric proxies accumulate to become surveillance
- ◇ Session data linked across venues becomes tracking data

The discipline: Define what is collected, how long it is retained, what is correlated, and when it is deleted. Audit the boundaries. Privacy is architecture, not policy.

Decentralized identity and verifiable credentials

The technical foundation for privacy-forward personalization already exists. Decentralized identifiers (DIDs) enable visitors to control their own identity without relying on venue databases. Verifiable credentials enable proof without disclosure—a visitor can prove they qualify for accessibility accommodations, senior pricing, or membership benefits without revealing their name, age, or medical status.

This inverts the traditional model. Instead of venues accumulating identity and visitors hoping it stays safe, visitors carry credentials, and venues verify only what they need to know.

This is the difference between proving eligibility and disclosing identity. Most identity checks are actually eligibility checks. The venue does not need your driver's license containing your name, address, or date of birth to make a decision. It needs a verifiable answer to a narrow question, such as "Over 21?", "Eligible for senior pricing?", "Has accessibility accommodation credential?", or "Has membership tier X?".

In this pattern, the venue sends a proof request for a single claim, and the visitor's wallet responds with a cryptographic proof that can be verified, without handing over the underlying document. This follows the standard verifiable credential exchange model: an issuer creates a credential, the holder stores it, and a verifier checks it.

The "Yes/No" Identity Handshake

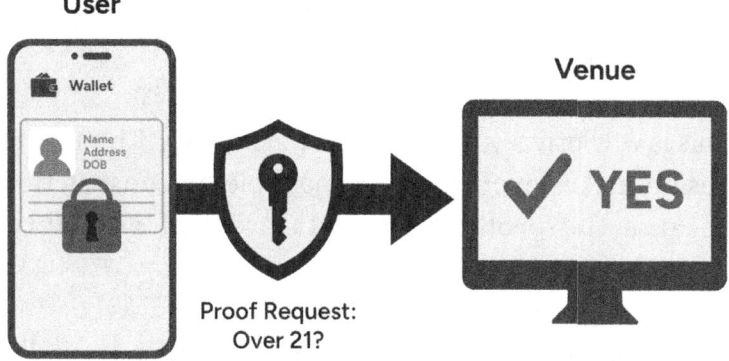

The venue receives the answer, not the document

Scan this QR code to view the figure on WorldModel.global

The important operational change is that the venue verifies only what it needs and does not, by default, accumulate a dossier.

What the venue receives, and what it must not receive

The venue should receive a yes/no decision and, if needed, minimal audit metadata (purpose, time, and outcome). It should not receive raw attributes (date of birth, address, ID number), full credential payloads, or anything that enables later correlation across visits. If the venue logs anything, it should log the decision, not the evidence.

This is why privacy is architecture, not policy. The system design makes oversharing impossible by construction because the protocol requests proof of eligibility, not identity.

Smart contracts can enforce these boundaries automatically. A credential exchange that was promised to be single-use can be architecturally guaranteed to remain single-use.

Consent given for one purpose cannot silently drift to another. The policy becomes the mechanism.

> ◆ *The guest should feel the venue is loyal to them, not extracting from them.*

The phone becomes an agent

The next shift is already emerging. The visitor's phone becomes an agent—software that holds preferences, holds credentials, negotiates what is shared, and answers venue requests on the visitor's behalf. Instead of apps everywhere demanding accounts, the phone becomes the stable layer and venues become interoperable requesters. The venue asks a question; the phone agent responds with minimal proof. This is not futuristic speculation—the standards exist today. For boards evaluating technology investments, this direction matters: systems designed around negotiated interactions will age better than systems designed around venue-controlled profiles.

For venues operating across jurisdictions with differing privacy regulations, or serving international visitors who expect their home-country protections, this architecture is not optional. It is the only scalable path to genuine compliance.

Anonymous-by-default preserves trust. Opt-in continuity creates value without eroding trust. Those who do not opt in should still benefit from a coherent day.

SCENE: OPT-IN CONTINUITY DONE RIGHT

A member opts in and receives benefits. A non-member does not opt in and still gets a coherent day. The system does not punish anonymity, and it does not build dossiers by default. Both paths work.

9

Virtual Docents and What They Must Never Do

Avatar guides become valuable when bounded and purposeful: controlled storytellers, not open-ended improvisers. Viewpoint makes experiences replayable without rebuilding physical space.

Imagine a museum tour led by an astronaut who explains a spacecraft as someone who lived inside it. On the next visit, choose a test pilot—same artifact, different story. The objects do not change. The lens does.

For brands, this is content leverage. For guests, a reason to return. For institutions, a way to grow meaning without tearing down walls.

What virtual docents must never do

- **Override safety constraints.** If a safety message must be delivered, the docent delivers it.
- **Require identity for basic help.** Guidance without logging in.
- **Infer sensitive attributes.** No guessing age, health, emotion, or politics unless explicitly provided.
- **Act as an unbounded author.** Draw only from the Body of Knowledge. No inventing facts.
- **Improvise facts.** If the answer is not in approved sources: "I don't have that information."
- **Contradict safety messaging.** In incident mode, safety-critical only.
- **Store conversations beyond the session.** Ephemeral by default. No surveillance of dialogue.

These constraints are not optional. They are what make conversational AI safe in public spaces.

10

Lifecycle Automation and Verification

Lifecycle automation prevents a venue from aging into a problem. Systems that delight on opening day drift into chronic friction when no one owns their long-term behavior under load.

The lifecycle loop

1. Occupancy detection with confidence levels
2. Power action staged by policy (audio first, then visual, then projection)
3. Pre-warm timing based on approach vectors
4. Verification that restoration actually occurred
5. Drift trending to catch degradation early
6. Preventive maintenance triggered by signals, not calendars

◆ *Commands are not outcomes.*

Verification sequence (non-negotiable)

1. Command issued (what was requested)
2. Action attempted (what the system tried to do)
3. Outcome observed (what actually happened in the venue)
4. Outcome evaluated (pass or fail against an acceptance test), with evidence recorded

If a vendor cannot prove outcomes, do not grant autonomy.

A projector can receive "on" and still show black. A display can be "on" and frozen. An audio zone can be "enabled" and silent.

Verification proves readiness—it does not assume it.

Drift signals to monitor

- Temperature trends rising over weeks
- Fan speeds increasing without matching workload
- Error rates creeping upward
- Output quality degrading slowly
- Network latency outside baseline
- Repeated "fixed by restart" patterns

Put these together, and the venue behaves like infrastructure: healthy, truthful, and responsive.

PART III

THE LADDER

11

Stage 0, Stage 1, Stage 2

The ladder provides a shared map so boards, designers, operators, and technology teams can argue productively about outcomes rather than terminology.

The Personalization Ladder

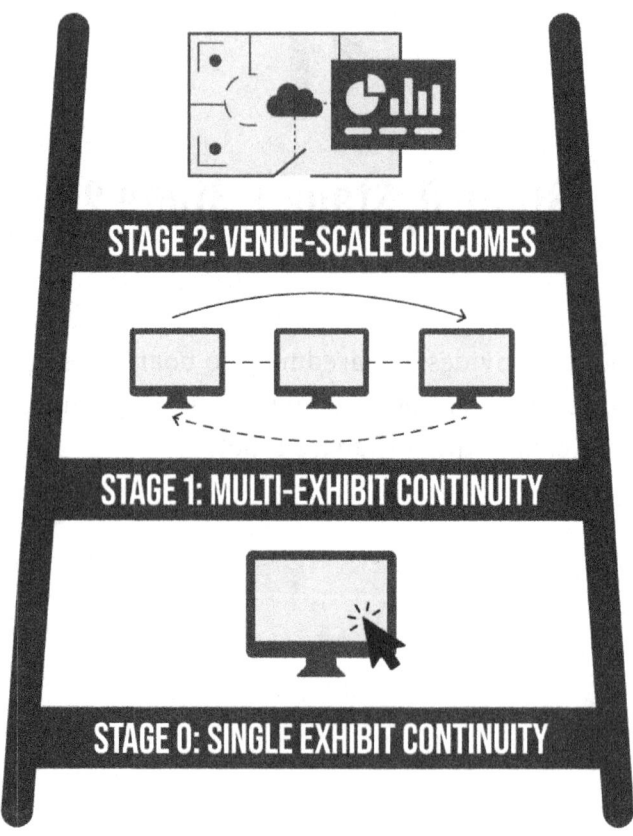

	SCOPE OF DECISION	PRIMARY RISKS	WHEN GOVERNANCE BECOMES MANDATORY
STAGE 2	Holistic, enterprise-wide strategy.	Systematic bias, major privacy breach.	Mandatory; formal policies, audits.
STAGE 1	Cross-exhibit, visitor journey tracking.	Data silos, disjointed narrative.	Emerging needs; basic standards.
STAGE 0	Localized, one interaction point.	Isolated failure, poor user experience.	Not typically required; ad-hoc.

Stage 0: Exhibit-scale personalization

Each exhibit operates independently. At each stop, visitors select language, depth, and accessibility.

You get: Language, depth, accessibility, privacy-forward recognition, virtual docents, Personal Channel, and offline-first reliability.

You do not get: Preferences that persist across exhibits.

Layers required: Layer 0 + Layer 1

Governance: Body of Knowledge constraints. No Cognitive Governance Layer™ (CGL™) required.

Stage 0 is a complete, valuable operating state. Many successful venues operate here.

Stage 1: Journey continuity

Preferences persist across multiple exhibits. Set language, age, depth mode, accessibility mode, and interests once; the venue honors them throughout.

Stage 1 is preference continuity, not profiling. Interests are declared, limited to venue-appropriate themes, and can be reset without friction. When age-related differences matter, treat them as visitor-chosen modes (family mode, student mode, calmer pacing), or as entitlements proven with minimal disclosure (for example, a verifiable credential for student or senior benefits), without collecting a birthdate.

You get: All Stage 0 capabilities plus preference persistence, anonymous session continuity, and a coherent multi-exhibit experience.

You do not get venue-scale coordination, flow balancing, or queue management.

Layers required: Layer 0 + Layer 1

Governance: Body of Knowledge constraints. No CGL required.

For most venues, Stage 1 strikes the right balance of capability and simplicity. Many venues should operate at this stage indefinitely.

Stage 2: Venue-scale orchestration

The venue operates as a coordinated system with a shared operational truth.

You get: All Stage 0/1 capabilities plus WorldModel™, venue-scale guidance, queue management, schedule-aware recommendations, cross-zone coordination, and incident-mode behavior.

You also get: governance requirements, a definition of the Constitution, CGL decision gates, increased complexity, and audit requirements.

Layers required: Layer 0 + Layer 1 + Layer 2 + Layer 3

Governance: CGL evaluates all venue-scale decisions against the Constitution before execution.

Stage 2 is not better than Stage 1. It is different. Proceed only when scale genuinely requires it.

Important: Stage 2 should be introduced by limiting scope, not by relaxing safeguards. Here, "scope" means the number of zones affected, the number of decision types the system is allowed to make, and the number of channels it controls.

Start smaller, but keep governance, verification, and consent boundaries intact from day one.

Stage selection (board shortcut)

Choose Stage 0 if: your goal is better interpretation, accessibility, and clarity without changing operations.

Choose Stage 1 if: you need personal delivery and inclusion at scale, but you do not need the venue to make consequential decisions.

Choose Stage 2 only if: your venue must coordinate routing, access, safety, and service dynamically, and you can fund governance, verification, and operations discipline as ongoing obligations.

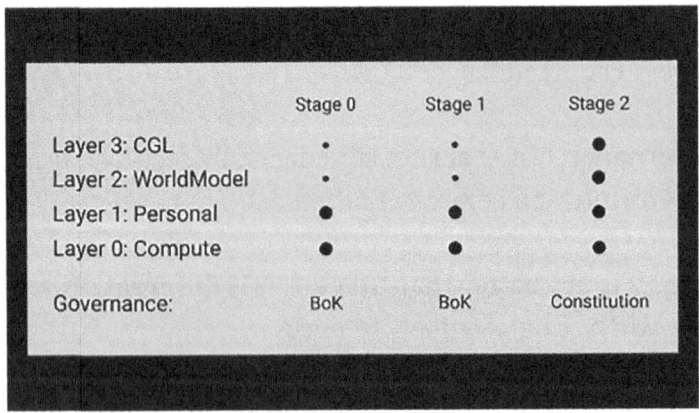

	Stage 0	Stage 1	Stage 2
Layer 3: CGL	·	·	●
Layer 2: WorldModel	·	·	●
Layer 1: Personal	●	●	●
Layer 0: Compute	●	●	●
Governance:	BoK	BoK	Constitution

12
Decision Guide: Which Stage is Right?

Question	Stage 0	Stage 1	Stage 2
Visitors move through multiple exhibits?	No	Yes	Yes
Repeating language selection is a problem?	No	Yes	Yes
Multiple systems must agree on venue reality?	No	No	Yes
Need flow balancing or queue management?	No	No	Yes
System makes decisions affecting routing or fairness?	No	No	Yes
Prepared to define and maintain a Constitution?	N/A	N/A	Required

The progression is additive

Need	Add	When necessary
Exhibit personalization	Base stack	Always - this is foundation
Journey continuity	Session state	Visitors cross multiple exhibits
Venue-scale coordination	WorldModel™	Multiple systems must agree
Consequential decisions	CGL + Constitution	Decisions affect flow/fairness/safety

What each stage requires

Stage	Layers	Governance
Stage 0	0 + 1	Body of Knowledge
Stage 1	0 + 1	Body of Knowledge
Stage 2	0 + 1 + 2 + 3	CGL, Constitution, Value System

PART IV

WORLDMODEL AND GOVERNANCE

13

The Discovery: When Personalization Meets Scale

The personalization stack is working and deployed. Visitors receive content in their language, at their preferred depth, on their own devices. Museums and brand centers operate successfully at Stages 0 and 1.

Then, clients ask for more.

A theme park wants personalization across an entire land. A cruise line wants continuity across a ship. A city wants coherence across a district.

The Discovery:

Exhibit-level intelligence does not scale to venue-level complexity.

Different systems hold different truths. Signage says one thing, apps say another, and staff say a third. The problem is not the personalization technology—it works. The problem is coordination across systems that do not share state.

That is when shared operational truth becomes necessary.

WorldModel™ is not a replacement for the personalization stack. It is an addition that enables venue-scale coherence.

But operational truth creates a new problem.

When a system can route visitors, balance queues, and shape flow, it can also harm. It can misroute. It can create unfair outcomes. It can optimize a metric while destroying the experience.

The concern: Capable AI without governance is dangerous in public spaces.

That is why CGL exists. It evaluates every proposed action against a Constitution that encodes values, hard constraints, and non-negotiable boundaries. The system cannot act unless governance permits.

Governance is not required at Stages 0 or 1. Body of Knowledge constraints address AI safety at the content level. CGL becomes necessary only when decisions affect flow, access, fairness, or safety at scale.

The four jobs of a living venue

A WorldModel Venue divides intelligence into four distinct responsibilities rather than running one monolithic "AI brain." This separation makes the system operable, auditable, and trustworthy.

- Memory (ICL™): Manages continuity, knowing the visitor in the lobby who asked for Spanish is the same visitor now in the gallery, without building a dossier. Enforces forgetting as a feature.
- Senses (EDE™): Models dynamics, not just snapshots but what is changing, how fast, and what happens next if the venue does nothing. Anticipates crowds, queues, and acoustic shifts.
- Coordinator (MAOL™): Proposes candidate actions when multiple plausible moves exist, reroute visitors, adjust signage, trigger staff notification, or shift comfort modes.
- Conscience (CGL™): Evaluates every proposed action and has the power to refuse. Enforces Constitutional rules at runtime, not merely in policy documents.

Keeping the conscience separate from the coordinator ensures that no matter how capable the system becomes, it cannot break the rules you set. In procurement, ask whether these responsibilities are architecturally separated or merged into an opaque decision engine.

14

WorldModel In Plain Language

WorldModel™ is a continuously updated representation of the venue as it is now: what things are, where they are, what state they are in, and what constraints apply.

Most venues already have models of reality. They just have too many, and they disagree with one another.

WorldModel contains

- ◇ Spatial truth: What exists, where, in what configuration
- ◇ Experience truth: What's playing, in what mode, in what language
- ◇ Operational truth: What's open, closed, available, constrained
- ◇ Acoustic truth: Where it's loud, quiet, where sound bleeds
- ◇ Equipment truth: What's healthy, degraded, failed
- ◇ Occupancy truth: Where people are, density, how it's changing
- ◇ Safety truth: Current posture, incident mode, constraints in effect

Shared WorldModel forces systems to stop debating. Decisions are made against a single, consistent view of reality.

WorldModel is required only at Stage 2. If your venue operates at Stage 0 or Stage 1, you do not need it.

How the World Model becomes WorldModel™

A governed architecture that transforms raw environmental intelligence into safe, value-aligned, human-centric operation.

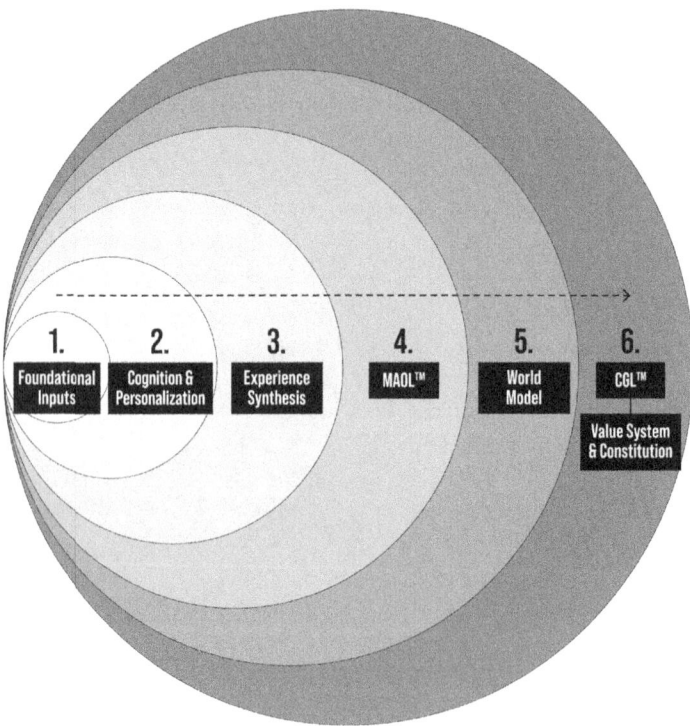

WorldModel™ is the complete, governed architecture that emerges when the World Model operates inside the Cognitive Governance Layer™

For the complete and fully detailed version of this image,
visit **worldmodel.global**

1. Turn raw data into meaningful context

- Edge compute nodes
- Venue sensor
  (light, sound, occupancy, flow)
- Recognition layer
- Object recognition
- System/Show control
- Content nodes & media assets
- Visitor inputs
  (opt-in identity, preferences, history)
- Personal devices (OnBoard layer, Personal Layer, Concierge layer)

2. Transform signals into adaptive responses

- Personalization layer
- Contextual intent inference
- Accessibility adaptations
- Learning loops
- Eligibility & access logic
- Avatar interactions
- Object and visitor recognition tie to Recognition layer output form foundational input

3. Generate dynamic, human centric actions

- Real-time content assembly
- Dynamic routing & Wayfinding
  (non-visual navigation layer)
- AR layer
- Adaptive storytelling
- Preference modeling
- Pattern recognition
- Cross-device session continuity

4. Coordinate all venue agents into a unified intelligence

- Multi-agent collaboration
- Cross-venue continuity
- Zone-to-zone synchronization
- Task allocation
- Conflict resolution
- Role switching
  (guide, concierge, expert)
- Agent states and behavior patterns

5. Structured representation of environment, guests, and state

- Environmental Dynamics Engine (EDE™)
- Environmental dynamics
- Prediction surfaces
- State histories
- Temporal Governance Framework (TGF™)
- Identity continuity (logical model)
- Venue map & topology
- Identity Continuity Layer (ICL™)
- Spatial & semantic relationships

6. Enforces values, safety, constraints, and accountability

- BehaviorGuardrails
- Safety Interlocks
- Role Governance
- Delegation Rules
- Constraint-Driven Decision Making
- Oversight & Explainability
- Audit Trails & Compliance
- Domain rules & constraints

15

Governance, Constitution, and Decision Gates

CGL™ is the layer through which consequential decisions must flow. It checks proposed actions against the Value System and Constitution, rejects violations, preserves the override capability, and records justification.

WorldModel is awareness. CGL is judgment.

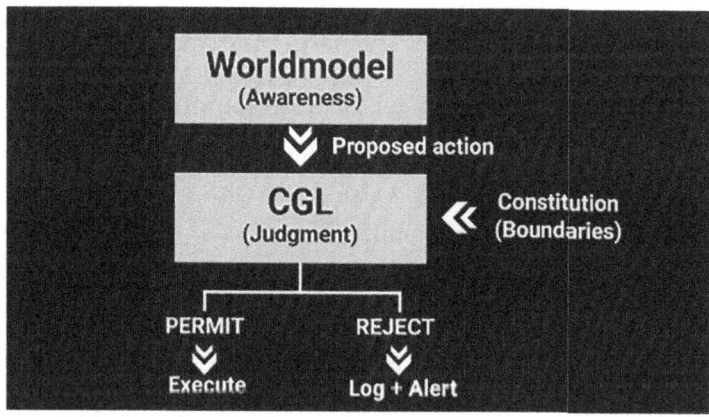

The Constitution

Every Stage 2 venue operates from a Constitution encoding:

Outcomes optimized:

- ◇ Visitor clarity and wayfinding
- ◇ Inclusion and accessibility
- ◇ Comfort and reduced fatigue
- ◇ Fairness in access and queuing

Hard constraints never violated:

- ◇ Safety requirements
- ◇ Accessibility obligations
- ◇ Privacy boundaries
- ◇ Fairness principles

Allowed actions by category:

- ◇ What the system can do autonomously
- ◇ What requires visitor consent
- ◇ What requires operator approval
- ◇ What is never permitted

Constraints are enforceable rules checked before execution and recorded for audit. A venue that can explain its actions can be trusted with autonomy.

◆ *If you want a venue that can think, you must also want a venue that can refuse to act.*

SCENE: THE SYSTEM REFUSES

A proposed re-route would reduce queues but would create an unfair access outcome. The decision gate rejects it, logs why, and falls back to a safe default. That is what governance means in practice.

Operating modes

◇ Normal: Full capability, standard constraints
◇ Busy: Tighter constraints, increased coordination
◇ Incident: Safety-critical only, non-essential suspended

Human override must be the norm. Operators override any automated decision immediately, without code changes.

Governance as executable contract

A Constitution is only as trustworthy as its enforcement. Smart contracts provide cryptographic guarantees that governance rules are followed—not merely promised, but enforced through the architecture.

When a system commits to discarding session data after 24 hours, that commitment can be verified, not merely asserted. When a venue promises that biometric data never leaves the premises, the architecture can make exfiltration technically impossible rather than merely policy-prohibited.

This matters for procurement. A vendor who claims governance but cannot demonstrate enforcement is offering policy. A vendor who can show you the contract logic is offering architecture.

PART V

WHY NOT

16

When This Approach Is Wrong

This framework is not universally correct. There are venues, circumstances, and strategic positions where a different approach is better. Intellectual honesty requires naming them.

When to choose differently

Your venue is genuinely simple.
A single-room gallery with one exhibit type and one audience segment does not need a personalization stack. A well-designed static experience, executed with care, may serve visitors better than technology that solves problems you don't have. Not every space needs to adapt. Some should simply excel at one thing.

Your budget forces a choice between foundation and finish.
If capital is constrained, spend it on the experience itself—better content, design, and staff training—rather than on infrastructure for personalization you cannot yet populate. A coherent but thin experience disappoints. Visitors forgive limited scope; they do not forgive broken promises. Stick with a Compute-node-based solution to safeguard your future, and limit it there.

Your organization cannot sustain the operating model.
The technologies described here require ongoing attention: content curation, Body of Knowledge maintenance, governance review, verification monitoring. If your organization lacks the appetite or staffing for sustained operational discipline, the system will drift into incoherence—the opposite of its purpose.

Your visitors do not want adaptation.
Some experiences are defined by their unchanging nature. A historic site, a memorial, or a place of worship may be more powerful for its constancy. Personalization implies that the visitor's preferences matter more than the place's intention. That is not always true.

Speed matters more than coherence.
A temporary activation, a pop-up experience, or a one-season installation may not justify the architectural investment. When the venue has a defined end date measured in months, simpler approaches—even fragmented ones—may be pragmatically appropriate.

Alternative approaches that work

Curated simplicity.
A venue that does one thing superbly, with no personalization, no adaptation, no recognition—just a single, carefully designed experience delivered consistently. This is harder than it sounds and, when achieved, is often more memorable than adaptive environments.

Human-first hospitality.
Staff-intensive service models in which trained humans provide continuity, translation, and adaptation. More expensive per visitor, but they create employment and a human connection that technology cannot replicate. For some brands and missions, this is the right choice.

Incremental evolution.
Rather than architectural transformation, continuous improvement of existing systems. Adding translation here, accessibility there, and better signage elsewhere. This approach is slower and may never achieve true coherence, but it is lower risk and requires less organizational commitment.

Platform dependence.
Accepting a major platform's ecosystem—a theme park management suite, a museum CMS, an airport operations platform—and operating within its constraints.

You trade flexibility for a reduced integration burden. For organizations without technical depth, this may be the responsible choice.

The honest trade-offs

The framework described in this book optimizes for venues that are large, long-lived, multilingual, accessibility-conscious, and operated by organizations with technical sophistication and governance discipline. It assumes the venue will evolve over a decade or more and that capabilities must compound rather than be rebuilt.

If those assumptions do not describe your situation, this may not be the right framework for you—although a Compute-node-based solution will still serve you well.

Choosing differently is not failure. It is clarity about what you are building, what you can sustain, and what you want visitors to actually notice. If your budget does not support Stage 2, incremental improvements are a valid path: add translation, expand accessibility, improve wayfinding, and tighten content quality until the experience is measurably better. But be honest about the ceiling. Coherence does not emerge from scattered upgrades. And if you decide to adopt advanced technology, do it with intent: fund the story, the content, the operations, and the governance first.

No one cares that you have the latest AI hardware in your museum if it does not support the story, or if you have nothing to show for it.

PART VI

DEPLOYMENT BY VERTICAL

17

Patterns Across Verticals

Across verticals, the architecture remains the same, but constraints and priorities differ.

These are pattern fits and diligence prompts based on operating constraints, not a claim of prior deployments by the author in every listed vertical.

Museums and cultural venues

Recommended: Stage 1. Stage 2 is recommended only for large multi-building campuses.

Curatorial intent is sacrosanct. The Body of Knowledge is essential—AI must preserve institutional voice. The win condition is clarity. If visitors must decide where to go, they stop thinking about what they came to see.

Operational reality: The failure mode is hesitation. Visitors stop, re-decide, and burn cognitive budget before they reach the artifact. That suppresses learning, pacing, and satisfaction.

What to demand:

- Body of Knowledge boundaries that preserve institutional voice under Q&A.
- Persistent language and accessibility across exhibits without re-asking.
- Degraded-mode behavior that preserves clarity, not just playback.

Theme parks and attractions

Recommended: Stage 2 for large parks and Stage 1 for smaller attractions.

The day is orchestrated. Better service with lower staffing requires coordination across the environment. Queue fairness becomes an explicit governance constraint.

Operational reality: The failure mode is unfairness and misrouting under pressure. Small contradictions become crowd waves, and staff heroics becomes the operating model.

What to demand:

- Constitution constraints that explicitly encode safety, accessibility, and fairness as non-negotiables.
- Outcome verification for routing, queue messaging, and incident-mode behavior.
- Clear boundaries on allowed actions, including refusal behavior when uncertain.

Cruise ships and resorts

Recommended: Stage 2.

Continuity is the business. Guests do not want a new operating system each day. Schedule density drives coordination. Coherence reduces cognitive load across dining, entertainment, and excursions.

Operational reality: The failure mode is fragmentation across days. Guests feel like they are learning a new operating system daily, and staff spend time re-explaining instead of serving.

What to demand:

- Journey continuity across dining, entertainment, excursions, and accessibility needs.
- Offline-first resilience across decks and dead zones.
- Clear retention and deletion rules for any continuity features.

Airports and transport hubs

Recommended: Stage 2.

Cannot hire your way through complexity: disruptions are too frequent and distributed. A delayed flight creates a wave; the venue immediately shifts guidance to the right language for accessible routes.

Operational reality: The failure mode is stress amplification during disruptions. A delay creates a cascade, and unclear guidance becomes missed connections and manual interventions.

What to demand:

- Real-time operational truth feeding guidance, with accessible routing as primary.
- Evidence that multilingual delivery stays coherent during incident and peak load.
- Operator override and audit trails for any automated re-routing.

Retail and mixed-use

Recommended: Stage 1 for most retail settings. Stage 2 for large destinations.

Consent is core. Relevance feels like service only when chosen. Recognition must be service-first, never surveillance. POS linkage is state-based, not profile-based—unless customers opt in and gain a benefit.

Operational reality: The failure mode is "relevance becomes surveillance." If recognition feels extractive, opt-in collapses, and brand trust is damaged.

What to demand:

- Anonymous-by-default operation that still improves wayfinding and service.
- Opt-in programs with explicit benefits and clear consent boundaries.
- Integrations that are state-based, not profile-based, unless customers explicitly opt in.

Corporate experience centers

Recommended: Stage 1 for most. Stage 2 for large, multi-zone facilities.

Governance is a feature. Decisions are auditable. Claims are consistent. The goal is credibility, not spectacle. Visitors who choose the deepest content (Student mode) are often the evaluators who decide whether to buy.

Operational reality: The failure mode is credibility loss. If the system contradicts itself, invents content, or fails under load, evaluators stop trusting the organization.

What to demand:

- BoK governance, plus explicit "must never do" constraints enforced in operation.
- Logs and verification proving readiness before live demos and events.
- Consistency across touchpoints, including staff tools and signage.

Districts and smart cities

Recommended: Stage 2.

Purest coherence problem. Every venue adds complexity. Civic context demands the strongest privacy: identity is rarely required for navigation.

Operational reality: The failure mode is privacy backlash and coordination collapse. Scale magnifies inconsistency, and identity requirements become unacceptable.

What to demand:

- WorldModel operational truth without identity as a prerequisite for navigation.
- Governance that constrains cross-venue correlation by design.
- Jurisdiction-aware privacy compliance implemented as architecture, not policy.

Universities, healthcare, public facilities

Recommended: Stage 1 for buildings. Stage 2 for large campuses.

High-stakes—visitors are not always relaxed. Clarity and inclusion are essential to dignity. When guidance is coherent, multilingual, and accessible by default, stress, errors, and interventions drop.

Operational reality: The failure mode is harm through confusion. Visitors are not relaxed, and mistakes have real consequences.

What to demand:

- Accessible guidance as the primary path, including degraded-mode behavior.
- Minimal-disclosure entitlements when needed, proven, not inferred.
- Auditability and refusal behavior for consequential decision automation.

PART VII

BOARD-LEVEL DILIGENCE

18

Why This Matters At Board Level

Boards do not govern architecture. They govern decisions, risk, and accountability. This section frames what the architecture enables, what it mitigates, and what to ask before approving it.

The strategic question

Not "should we invest in technology?" Every venue does. The question is whether investment compounds or fragments.

Fragmented investment buys screens, apps, and systems that solve local problems while creating global incoherence. Each addition makes the next harder.

Compounding investment builds on a substrate that enables growth without reconstruction. Language expands without reprinting.

Accessibility improves without retrofit. The venue becomes more capable each season.

Three levels of commitment

Stage 0: Remove black-box dead ends. Move to a Compute-node posture. Prove the substrate. Minimum commitment to prevent future lock-in.

Stage 1: Persistent language, depth, accessibility, and interest continuity across exhibits. Anonymous session state. Where most venues should aim, and many should stop.

Stage 2: Shared operational truth. Cross-zone coordination. Consequential decisions constrained by the Constitution. Only for theme parks, ships, airports, and districts where scale demands it.

Questions for the board

1. Which stage are we targeting, and why?
2. What is the default privacy posture? Is identity required for basic operations?
3. What happens when we want more capability next year—upgrade or rebuild?
4. What evidence exists after a bad day?

To delegate effectively

Ask the COO/CTO to review the reference standard and return with:

1. Recommended stage target with justification
2. Minimum architecture and governance requirements for procurement
3. Stage 0 or Stage 1 pilot plan with evidence gates, stop rules, rollback, and a full operating-cycle test.
4. Stage 2 implementation plan using bounded scope: limited zones, limited decision types, and limited touchpoints, with full governance and full outcome verification from day one.

If verification is missing, the system is not autonomous. It is guesswork.

The economic frame

Conservative ranges from published research:

◇ Revenue uplift 2–7%, profit uplift 1–2%
◇ Aggressive: revenue uplift 10–15%, cost-to-serve reduction 15–20%

Outcomes depend on execution, not on technology. A well-operated Stage 1 outperforms a poorly executed and operated Stage 2.

The durable argument is defensive: venues without coherence pay a catch-up premium later.

19

What To Demand In Procurement

Buy outcomes and evidence, not features.
Features invite debate. Outcomes demand proof.

Ask for non-proprietary hardware and upgrade paths.
Bespoke appliances create a lifecycle trap.

Ask where shared state lives, where decisions are made, and how actions are verified.

If the vendor cannot explain graceful degradation, the system will simply stop working when something fails.

Demand a privacy posture:
anonymous by default, opt-in by design.

Demand an IDEA and ESG baseline, not just a feature list. Require multilingual delivery, hearing-aid compatible personal audio, captions, sign language options where provided, Braille where physical labeling exists, and non-visual navigation support for blind and low-vision guests. Require lifecycle clarity: published support life, modular replacement strategy, update policy, security posture, and measurable performance under peak load.

Ask about verifiable credentials and smart contract enforcement.
Can visitors prove entitlements without surrendering their identity? Are governance commitments policy-based or enforced through the architecture? As privacy regulations diverge across jurisdictions, venues need systems that can satisfy multiple regimes simultaneously—not through legal workarounds but through technical architecture that makes violations impossible.

Demand evidence:
what is logged, what is provable after a bad day, and how audits are satisfied.

Ask about three modes of AI delivery.
Cannot distinguish between curated, BoK, and hybrid? Cannot explain what AI is NOT allowed to do? Buying risk.

Ask about offline-first. Critical paths in the cloud = fragility at peak.

Require stage clarity. Ensure the vendor understands and delivers at the needed stage, not selling Stage 2 complexity when Stage 1 serves.

What not to buy (fast filters)

- A system that only works with identity, login, or an app.
- A system that cannot explain its decisions with evidence and logs.
- A system that demos well but cannot state performance under load and degraded conditions.
- A system that treats accessibility as an add-on rather than a default delivery mode.

20

What To Demand In Operations

Demand operability, graceful degradation, and predictable recovery.

Operability: staff see state, understand changes, and safely override. Degradation: failures are fail-safe. Recovery: service restores without heroics.

Track intervention load: staff interventions per hour during peak. Rising intervention load indicates growing fragmentation.

Demand verification, not commands. System proves zones are ready, not merely claims commands were sent.

Questions after an incident

- ◇ What changed, when, and why?
- ◇ Which rule allowed it?
- ◇ Which data was used and discarded?
- ◇ Which override applied?
- ◇ What evidence exists?
- ◇ What did verification confirm?
- ◇ Were accessibility constraints maintained?
- ◇ Was the Constitution violated?

If you cannot answer these questions, you have a toolkit, not an operating model.

Content operations is an ongoing discipline

Personalization is not a one-time installation. It creates a publishing surface. Language variants, depth layers, accessibility formats, and Body of Knowledge curation require ongoing attention—versioning, approvals, staging, and rollback capability.

A venue that treats content as a static asset will drift into incoherence. A venue that treats content as behavior—the outward expression of operational truth—will remain alive.

In procurement, ask: Who owns the content publishing workflow? What is the approval cadence? What happens when a correction is needed? If the answer is "call the vendor," you are renting capability, not building it.

21

Questions For Partnership Conversations

Hardware independence

- ◇ Replace failed hardware with non-proprietary alternatives?
- ◇ Documented substitution strategy for supply chain disruptions?
- ◇ Standard or proprietary interfaces?

Software control

- ◇ Do we own the configuration or does the vendor?
- ◇ Can we export the complete system state?
- ◇ Can another integrator service this?

Data sovereignty

- ◇ What data is retained? How long?
- ◇ Where stored? How deleted?
- ◇ What happens if the relationship terminates?

Upgrade path

- ◇ Capabilities added without hardware replacement?
- ◇ Incremental or wholesale only?
- ◇ Expected upgrade cost over ten years?

Unsatisfactory answers = renting access to the vendor's platform, not buying a system.

PART VIII

THE CATCH-UP PREMIUM

22

What Happens If You Do Nothing

Doing nothing is still a plan: a plan to pay the catch-up premium.

You will still spend, but on patches, integrations, and staffing to cover seams. Meanwhile, a competitor builds coherence. Guests feel it quickly. They describe it as easier, calmer, and more personal.

That description travels further than your campaigns.

The next decade will see the construction of entertainment destinations, cultural districts, and smart cities at scales never before attempted, many in regions building entirely new visitor economies. These developments cannot rely on retrofitting legacy infrastructure. They require coherence from the ground up.

◆ *Visitors remember how the day felt.*

Operators who move first define what the market expects. Operators who move later pay to meet that expectation after it becomes a baseline.

Moving with intent is cheaper. Build the substrate. Prove Stage 1. Keep anonymity as the default. Then, let benefits compound season after season.

If Stage 2 becomes necessary, the foundation will be ready. If not, you will have avoided unnecessary complexity.

Final thesis

Coherence is a competitive position because it is an operating architecture. The winners will not be those who install the most technology. They will be those who eliminate contradictions, protect trust, and keep inclusion true under load.

Board next steps

- Pick the target stage, and write the non-negotiables: privacy posture, IDEA baseline, and verification requirements.
- Require evidence, not claims: measurable performance under load, safe failure behavior, and auditable logs.
- Scale only what you can govern and verify. If Stage 2 is required, start with bounded scope, not relaxed safeguards.

APPENDIX A

Executive Glossary

Anonymous operation: Default posture using a non-identifying state, discarding what is not needed.

Black box: Proprietary appliance with no or limited replacement path. Anti-pattern.

Body of Knowledge (BoK): Institution-controlled source material that bounds AI generation.

CGL™ (Cognitive Governance Layer™): Decision gate that evaluates proposed actions against the Constitution before execution. Stage 2 only.

Coherence: Venue behaves as a single place. No repeated context rebuilding.

Constitution: Explicit non-negotiable rules. The laws the system must abide by. Stage 2 only.

Curated/BoK/Hybrid delivery: Three modes that control AI generation boundaries.

Decentralized Identifier (DID): Visitor-controlled identity that does not depend on venue databases. Enables proof without disclosure.

Drift: Slow divergence from intended behavior in equipment or a system.

EDE (Environmental Dynamics Engine): Models venue dynamics—what is changing, how fast, and what happens next if nothing changes.

ESG: Environmental, Social, and Governance treated as measurable operating constraints, including energy, Compute, lifecycle serviceability, and auditability.

Experience debt: Accumulated incoherence from systems that each model reality differently. Shows up when operational promises disagree.

Governance: Enforceable decision gate. Stage 2 only.

ICL (Identity Continuity Layer): Manages session continuity and enforces forgetting. Prevents identity from becoming accidental.

IDEA: Inclusivity, Diversity, Equity, and Accessibility treated as acceptance criteria and degraded-mode requirements, not accommodations.

Intervention load: Staff interventions per hour. A proxy for fragmentation.

MAOL (Multi-Agent Orchestration Layer): Generates and coordinates candidate actions across competing objectives for governance evaluation.

Node-based Compute: Replaceable general-purpose nodes. Upgrade by provisioning.

Offline-first: Critical paths run locally without cloud dependencies.

Outcome verification: Confirms that actions took effect. Commands do not equal outcomes.

Personal Channel: Private delivery layer sending multilingual audio, captions, sign language, and accessibility formats to visitor devices.

Phone agent: Software on a visitor's device that holds preferences, holds credentials, and negotiates with venues on the visitor's behalf.

Programmable Canvas: Software-defined display surface that can change language, content, and function without physical modification.

Smart Contract: Self-executing governance logic that enforces commitments architecturally rather than through policy.

·**Stage 0 / Stage 1 / Stage 2:** Personalization maturity levels. Stage 0: exhibit-scale. Stage 1: journey continuity across exhibits. Stage 2: venue-scale orchestration requiring governance.

Streaker/Stroller/Student: Depth modes. Fifteen seconds / one or two minutes / unlimited.

Value System: Priority resolution. Safety over throughput.

Verifiable Credential: Cryptographic proof of entitlement (membership, accessibility, age) without revealing underlying identity.

Visitor mode: Declared experience mode (family mode, student mode, comfort mode) used for pacing, depth, and delivery without profiling.

WorldModel: Shared operational truth. Stage 2 only.

Sources

[1] McKinsey & Company, «Best of both worlds: Customer experience for more revenues and lower costs,» April 2014. https://www.mckinsey.com/capabilities/growth-marketing-and-sales/our-insights/best-of-both-worlds-customer-experience-for-more-revenues-and-lower-costs

[2] McKinsey & Company, «Prediction: The future of CX,» February 2021. https://www.mckinsey.com/capabilities/growth-marketing-and-sales/our-insights/prediction-the-future-of-cx

[3] Reichheld, Frederick F., *The Loyalty Effect: The Hidden Force Behind Growth, Profits, and Lasting Value*, Bain & Company, 1996. https://www.bain.com/insights/books/the-loyalty-effect/

[4] Arival, "Attraction Wait Times Skewer Perceived Value," 2020. https://arival.travel/article/attraction-waiting-times-skewer-perceived-value/

5. National Institute of Standards and Technology, "Artificial Intelligence Risk Management Framework (AI RMF 1.0)," NIST AI 100-1, January 2023. https://www.nist.gov/itl/ai-risk-management-framework

6. World Wide Web Consortium (W3C), "Verifiable Credentials Data Model v2.0," W3C Recommendation, May 2025. https://www.w3.org/TR/vc-data-model-2.0/

◆

Companion Resources

The World Model:

Governed AI and Hyper-Personalization in Physical Spaces

The complete reference standard includes full architecture specifications, governance patterns, procurement language, and acceptance criteria.

Detailed diagrams, downloadable procurement checklists, stage assessment tools, and updates to the framework are available at:

worldmodel.global

The World Model
Complete reference standard with architecture specifications, governance patterns, and procurement criteria

◆

A Note from The Author

This book exists because of a frustration that became an obsession.

For years, I watched venues struggle with the same problems. A proprietary box would reach end-of-life. Spares would vanish. A "small upgrade" would cascade into a capital project. Systems that worked beautifully on opening day would drift into chronic friction because no one owned their long-term behavior. Everywhere, the same pattern: local solutions creating global incoherence.

The breakthrough came when we stopped accepting that pattern as inevitable.

We began building systems on commodity Compute—general-purpose nodes that could be provisioned, replaced, and upgraded without forklift upgrades. As soon as that architecture was in place, doors opened. If a node could run software, it could support any number of languages. It could generate spatial audio environments. It could synchronize across devices. It could run AI models on-site, without cloud dependency—if not immediately to meet all requirements, then certainly soon, without significant changes.

Each capability led to the next, and each revealed possibilities we hadn't anticipated.

Then came the harder questions.

Once you can recognize visitors, how do you protect their privacy? We concluded the answer was architectural: keep everything local, require no internet connection, encrypt by default, and discard what isn't needed. Privacy couldn't be a policy bolted on afterward. It had to be a design constraint from the start.

Once you can personalize delivery, how do you ensure inclusion? We built systems for visitors who are blind, who use hearing aids, who need sign language, and who speak languages the venue never planned for. Accessibility stopped being an accommodation and became a default capability.

Once you can coordinate across an entire venue, how do you prevent the system from causing harm? A system capable of routing visitors and balancing queues can also create unfair outcomes, optimizing metrics while destroying experiences. That realization led to governance: a Value System that encodes priorities, a Constitution that defines boundaries, and decision gates that prevent action unless rules permit it.

I wrote this book because the industry needs a shared vocabulary for these capabilities, not tied to any particular implementation, but grounded in the problems implementations must solve. Whether you build these capabilities internally, work with specialists, or pursue a different path entirely, the questions remain the same: Which stage is right for your venue? What privacy posture will you maintain? What governance will you require?

My hope is that this framework helps you think more clearly about those questions. The reference standard, named *The World Model, Governed AI for Hyper-Personalized Venues* provides the technical depth. The companion website offers tools and updates. But the thinking starts here, with the recognition that coherence is not a feature—it is an architecture, a discipline, and ultimately a competitive position.

If any of this resonates with the challenges you're facing, I'd welcome the conversation.

Maris J. Ensing, February 2026 maris@worldmodel.global

About the Author

Maris J. Ensing is the founder of Mad Systems and a technology executive with issued patents and active patent applications in hyper-personalized media delivery, visitor recognition, and intelligent venue systems across the United States, Europe, China, Hong Kong, the Kingdom of Saudi Arabia, and the United Arab Emirates.

Based in Orange, California, his work focuses on AI governance, privacy-forward recognition, and governed AI frameworks for audiovisual and interactive physical environments at scale.

For speaking invitations, interviews, or professional correspondence: **maris@worldmodel.global**.

Other References

A1. EY, "How digital twin technology and AI can reimagine theme park experiences," Sep 4, 2025. https://www.ey.com/en_us/industries/media-entertainment/unleashing-theme-park-technology-transformation

A2. NIST, "Artificial Intelligence Risk Management Framework (AI RMF 1.0)," NIST AI 100-1, Jan 2023 (PDF). https://nvlpubs.nist.gov/nistpubs/ai/nist.ai.100-1.pdf

A3. NIST, "Artificial Intelligence Risk Management Framework: Generative AI Profile," NIST AI 600-1, 2024 (PDF). https://nvlpubs.nist.gov/nistpubs/ai/NIST.AI.600-1.pdf

A4. ISO/IEC, "ISO/IEC 42001:2023 - Artificial Intelligence Management System (AIMS)." https://www.iso.org/standard/42001

A5. W3C, "Decentralized Identifiers (DIDs) v1.0," W3C Recommendation, Jul 19, 2022. https://www.w3.org/TR/did-core/

A6. W3C, "Verifiable Credentials Data Model v2.0," W3C Recommendation, May 15, 2025. https://www.w3.org/TR/vc-data-model-2.0/

A7. NIST, "Digital Identity Guidelines (SP 800-63-4)," current revision landing pages and drafts. https://pages.nist.gov/800-63-4/

A8. W3C, "Web Content Accessibility Guidelines (WCAG) 2.2." https://www.w3.org/TR/WCAG22/

A9. NIST, "Privacy Framework: A Tool for Improving Privacy through Enterprise Risk Management, Version 1.0," Jan 16, 2020 (PDF). https://nvlpubs.nist.gov/nistpubs/CSWP/NIST.CSWP.01162020.pdf

A10. NIST, "Face Recognition Vendor Test (FRVT), Part 3: Demographic Effects," NISTIR 8280, 2019 (PDF). https://nvlpubs.nist.gov/nistpubs/ir/2019/nist.ir.8280.pdf

A11. Bluetooth SIG, "An Overview of Auracast™ Broadcast Audio," May 14, 2024 (PDF). https://www.bluetooth.com/wp-content/uploads/2024/05/2403_Auracast_Overview.pdf

A12. IEC, "IEC 60268-16:2020 - Sound System Equipment – Part 16: Objective Rating of Speech Intelligibility (STI)." https://webstore.iec.ch/en/publication/26771

Made in the USA
Coppell, TX
23 February 2026

72263329R00079